ABOVE: Nowadays the millwright can call upon modern machinery to assist him in his task. In 1976 a mobile crane was used to help remove the old sails of Thorpeness post mill in Suffolk.
OPPOSITE: A stone mill for grinding wheat, etc.

MILLS AND MILLWRIGHTING

John Vince

Shire Publications Ltd

CONTENTS

Printed in Great Britain by C. I. Thomas & Sons (Haverfordwest) Ltd, Press Buildings, Merlins Bridge, Haverfordwest.

ACKNOWLEDGEMENTS

The author records his thanks to the following: Finch Foundry Trust, Hereford and Worcester County Museum, the Chiltern Society, Chris Wallis and George Cusack. Illustrations are acknowledged as follows: Lewis Cameras, Melton Mowbray, page 24 (left); Museum of English Rural Life, Reading University, pages 5, 6, 8 (bottom), 12, 13, 17, 18 (bottom), 19 (left), 31, 32 (top), inside front cover; D. Nicholls, pages 1, 24(right), 26, 27, 29, 30, 32 (bottom); Sotheby & Co, pages 22, 23. Other photographs are by the author.

OPPOSITE: In a quiet corner of a country watermill we see a mill thrift (the handle which held the mill bill, the tool used for dressing the stones), a spare blank cog—and a reminder that stone dressing is thirsty work.
BELOW: Both windmills and watermills make use of a wheel called the wallower to transfer the drive to the main shaft. The oldest wooden wallower in England belongs to the seventeenth-century Lacey Green windmill in Buckinghamshire. This type of wheel is called a compass arm wheel because its spokes radiate from the main shaft in four directions.

INTRODUCTION

Water and wind power were almost the only means of grinding corn mechanically until the nineteenth century. Water power, which had other uses as well, survived into the industrial revolution and after; wind power, never so reliable, was not adapted for other uses except that of raising water and has scarcely survived at all, although environmentalists hope that it will be able to generate electricity for homes.

The Romans are thought to have built watermills in Britain although the first documentary record is of a Saxon mill in Kent in AD 726. The familiar design, first described in the first century BC by the Roman engineer Vitruvius, continued in general use as watermills spread through the country wherever there was a suitable flow of water. The wheels had horizontal shafts with flat buckets that dipped into the fast-running stream and were pushed round on the undershot principle. By the time of the Domesday survey in 1086 there were 5624 watermills in England.

The Crusaders are said to have introduced the windmill to Britain, although the post mill may have originated here or elsewhere in Europe. Early windmills were probably quite small structures with one pair of stones, but remarkable for the ingenuity of their design. Developments in design would have been gradual and the early improvements were not recorded: the technical details did not concern the unknown artists who decorated the borders of works such as the Lutterell Psalter, though their representations of early post mills are nevertheless useful.

The men who designed and built these early machines and their successors, which were so important to daily life, remain anonymous. Inventions made in the industrial era were documented, and from the eighteenth century onwards we know much about technical developments. But it was the work of the medieval engineers and the skills they developed which formed the foundation of the technical excellence which blossomed in later Georgian England.

This book shows how mills worked and how wide in scope was the work of the millwright, the first mechanical engineering profession; it also describes some of the splendid restoration work now being done by both amateur and professional millwrights. Recently there has been a renewed interest in mills and many have been saved from total decay by enthusiasts who have devoted years of effort to such projects. Some mills have been restored to working order and now enjoy a new lease of life.

HAND MILLS

The technique of grinding grain between two stones (querns) evolved in prehistoric times. At first the saddle-stone method was used: a small stone was rubbed backwards and forwards on a large block. Then an important development was made: one round stone was revolved over another of equal diameter. A small hole near the edge of the upper stone allowed a stick to be inserted, forming a handle with which the operator could quite easily rotate the upper stone. Grain placed at the centre of the stones was expelled around their circumferences. This quern method of reducing grain to flour or meal remained the same, whatever the motive power, until roller mills came into use during the Victorian era. The aim was always to produce the best possible flour. The application of mechanics increased the output of flour but the machine had to be large.

Using the energy produced by windmill sails and waterwheel was one of man's greatest inspirations. Mastery of this technique enabled later engineers to evolve designs incorporating the bevel, cycloid and other complex gears of modern technology: the importance of changing motion from one plane to another was a crucial first step in mechanical engineering.

No machines constructed by medieval engineers have survived to modern times. We can nevertheless deduce the methods they used by studying the features of surviving hand grinding machines that date from the seventeenth century. Grinding machines like the examples shown here would have been used in large households and were probably important status symbols in the kitchens of the day. Hand mills are certainly ancient in origin. In the first century BC Virgil referred to mills turned by hand. A *manimola* is mentioned in a document concerning Walton-le-Soken, Essex, in the twelfth century. At Wickham St Paul, in 1292, a *mola manualis* was recorded. It is unlikely that the design of the hand mill altered significantly during the middle ages.

Even the small hand-operated mill presented the millwright with the same kinds of problem as those he encountered in large watermills and windmills. The captions to the pictures explain some of the technical features.

OPPOSITE PAGE, TOP: A hand-operated mill (dated 1608), which demonstrates clearly how a vertical motion could be translated into a horizontal plane with two simple gear wheels. The framework of the machine is made of oak to provide a secure foundation for the moving parts. At each side of the frame the spindle 1, on which the vertical wheel 2 is mounted, projects to enable handles to be fixed. This wheel is made from solid members joined together with metal plates and an iron strip is nailed around the wheel's edge. On its inner face the wheel has square wooden teeth (several of which are missing), which made contact with the second gear wheel 3, parts of which also are missing. Its construction is quite different from that of the first wheel: two wooden discs were joined by a series of wooden staves around its edge, giving the wheel the appearance of a medieval horn lantern. This is why primitive wheels of this kind are called lantern wheels. When the spindle was rotated the teeth of the vertical wheel came into contact with the staves of the lantern and turned them round.

BOTTOM: The spindle of the lantern wheel was fixed to the upper runner stone, which rotated at the same speed. To allow the lantern wheel and the runner stone to be adjusted, the beam supporting the stone spindle 4 could be realigned with wedges. If the lantern wheel and stone had to be lowered, the left-hand wedge 5 was released and the other wedge 6 tapped down. With an arrangement of this kind very precise adjustments could be made to the distance between the two stones. Without the advantage of precision tools early millwrights could achieve considerable degrees of accuracy. The round holes at the edge of the solid wheel mark the position of the square-ended teeth. There was a good reason for round holes instead of the usual square mortice: a round housing allowed each tooth to be adjusted to compensate for wear.

ABOVE: With the hopper removed from the hand mill we can see how the component at the top of the vertical spindle was fashioned to support the runner stone. The cross-shaped iron is called the mill-rynd or mace. The furrows on the face of the stone helped to expel the ground meal around the stone's circumference. The hooks fixed to the framework on each side of the meal spout were provided to support a sack.

LEFT: Detail of a later hand mill showing how the teeth of the lantern wheel mesh with the cogs of the solid wheel.

This small wheel, less than 4 feet (1.22 metres) in diameter, is used to work a double pump. In its restored condition it can be seen at the Finch Foundry Museum, Sticklepath, Devon. Wheels of this kind had many uses, including the circulation of water in ornamental gardens.

WATERWHEELS

Roman and Saxon waterwheels were made of wood, the only suitable material for the purpose until the discovery of cast iron a millennium later. Elm is one of the best timbers to use for machinery which is constantly wet. There are limitations to the manner in which a wooden wheel can be constructed and the design of waterwheels altered very little through the centuries. The heaviest part of a wooden wheel was its axletree. This, as the name suggests, was made from the stock of a tree. Spokes radiated from the axle and supported the wheel's rim. The rim housed the paddles that beat the water as the wheel turned.

In the eighteenth century wood gave way to cast iron, whose durability and strength allowed founders to produce smaller and lighter components. The change from timber to cast iron was slow — some wheels were constructed with both iron and timber parts — but when wheels had to be replaced iron was used. Each wheel presented its individual problems. When a wheelhouse had been built around a wheel it might be difficult, through lack of space, to remove damaged parts. Since the largest part of the wheel is the axletree, wheelhouses were usually made with a door in line with the axle so that it could be removed if necessary.

The flow of water to the wheel had to be regulated, in most situations, by sluices, bypasses and millponds, according to whether the supply needed to be restricted or accumulated. Waterwheels were of three kinds: undershot, where the water passed under the axle — the simplest but least efficient method; breastshot, with the water at axle level — nearly twice as efficient; and overshot, in which the water passed over the wheel — slightly more efficient still.

Eighteenth-century millwrights had to master the use of cast iron as well as traditional timber; present-day millwrights need a sound knowledge of new materials such as wood preservatives and resin-based adhesives. The problems millwrights encounter today can sometimes be resolved by using techniques that had not been invented half a century ago: millwrighting these days is a mixture of tradition and new technology.

TOP: A large iron undershot wheel at Pann Mill, High Wycombe, Buckinghamshire.

LEFT: During the Second World War many mills were reinstated to aid the war effort. This photograph shows the wheel of Aymestrey Mill, Hereford, being rebuilt in 1940. Some mill wheels were also adapted to drive generators to supply lighting, so that the miller could work after dark.

OPPOSITE: The completed wheel at Mapledurham Mill on the Thames near Reading was installed in 1977. It is 15 feet 6 inches (4.72 metres) in diameter and 4 feet 5 inches (1.35 metres) wide. It generates about 20 to 22 horsepower (15 to 16 kilowatts).

OPPOSITE TOP: A water supply was sometimes carried to a wheel by way of a launder like this one at the Finch Foundry, Sticklepath, Devon. Several wheels were powered from it. Notice how close together are the legs which carry the weight of water in the trough. The beams joining the legs are wider than the trough to allow short buttresses to be fitted to strengthen the sides and resist the outward thrust of the water.

ABOVE: A Devonshire sluice gate which is controlled by chains wound round two solid wooden drums.

BELOW: Some gates were operated with iron gears. As the shaft A was rotated, the rack B was lifted or lowered.

FAR LEFT: One of the let-offs on the Sticklepath launder. A guillotine gate controls the supply of water to the overshot waterwheel. This iron wheel, by Pearce of Tavistock, has wooden paddles.

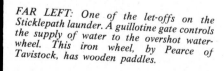

NEAR LEFT: Some wheels had spur wheels attached to the axle. These were used to supply power direct to the mill's ancillary machinery.

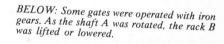

11

Watermill machinery. The waterwheel's axle A is shared by the pit wheel B, part of which is below floor level in a pit — hence the name. The vertical main shaft S is positioned above the axletree and at its lower end is the wallower W, turned by the pit wheel. The spur wheel P, above the wallower, carries the motion to the stone nuts.

WATERMILL MACHINERY

Although each mill had its own individual characteristics, the method of driving the machinery followed a common pattern. The vertical movement of the waterwheel was duplicated inside the mill by the large *pit wheel* on the same axle. This meshed with the horizontal *wallower* which turned the vertical *main shaft*. Above the wallower the great *spur wheel* on the main shaft meshed with the *stone nut*, through which passed the shaft which turned the stones. Near the top of this is the *crown wheel*, which turned the *lay shaft* by means of a bevel gear. The lay shaft carried various pulleys which operated ancillary machinery such as the sack hoist.

TOP: Millwrights had mastered the elements of mechanics in the middle ages. This view of the gears in a Berkshire Watermill, reveals how the speed of the waterwheel was increased by the pit wheel. Each turn of the pit wheel caused the wallower, with fewer teeth, to rotate several times. In its turn the wallower turned the shaft. The great spur wheel, with a larger diameter and consequently more teeth, increased the speed again when it came into contact with the stone nut N. The spindle on which the stone nut was mounted turned the upper runner stone, which was situated on the floor above. A spur wheel usually drove two stone nuts. The framework F had to be heavy in order to support the weight of the stones on the floor above and the inner end of the exletree. The preparation of the timber was a time-consuming task in the days before power saws. Most of the timber in old mills was cut out by hand over a dusty sawpit.

This pit wheel remains in position although the rest of the mill has been demolished. Like most pit wheels it was made in two halves so that it could be more easily fitted in the confined space between the vertical shaft and the pit wall. In Queen Victoria's reign many old wooden pit wheels were replaced by iron ones. Even if there had been sufficient space inside the mill, it would have been impossible to fit a single-piece wheel without removing the waterwheel's axle. The octagonal centre suggests that the pit wheel was made to fit an original wooden axletree. The present iron axle could be of a much later date than the pit wheel itself.

At the end of the day's work the stone nuts were usually raised up and put 'out of gear' so that they could not operate accidentally if the wheel started up. Wheels could 'run away' if heavy rain caused the water to run over the sluice or the bypass stream became blocked. Most mills used a simple screw device to raise the stone nut on its spindle. A jack ring J was placed below the wheel and when the screw was wound up the two rods R pushed the ring upwards, lifting the wheel above the level of the spur wheel's teeth. In this photograph the stone nut and spindle have been removed for repair.

BELOW: A sack-hoist chain with its well-worn barrel.

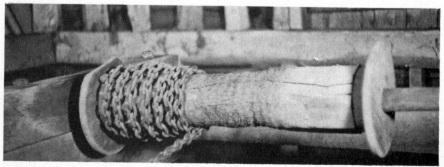

ABOVE: Power to drive the ancillary machinery was taken from the top of the main shaft. The upward facing teeth of a crown wheel C turned a bevel gear (missing in this photograph but positioned at D), which operated a layshaft L that contained a number of pulleys P. Belts from each pulley provided motion for such things as the sack hoist, grindstone, flour dresser and bolter.

RIGHT: Fixing teeth to a crown wheel.

ABOVE: When the wheat meal had been ground by the stones it had to be left for several days to cool before being passed through the dresser to separate flour from bran. When cooled, it was easier to separate the bran particles and refine the wheat meal so that it became millstone flour. This process enabled the bud of the wheat berry (the wheat germ) to survive. Dressers could be driven by belts or cogs.

BELOW: A bolter was used to separate different grades of flour. Like the dresser, it was driven by belts or gears but it revolved at a much slower rate. A seamless bolting cloth was stretched along its six arms. As they rotated the cloth billowed out and brushed against one of the fixed longitudinal bars B. This caused the flour to pass through the cloth, fine flour at the top, coarser flour lower down.

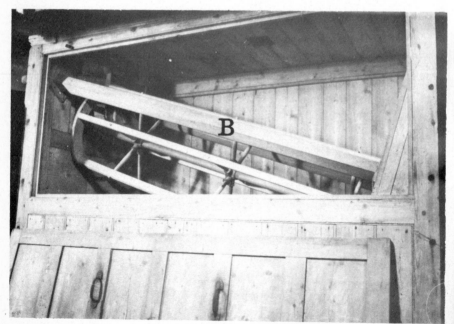

Each pair of stones was protected by a wooden casing or vat C, which provided a platform for the horse H. The horse supported the hopper P. The chute E carried the grain from the bin floor above into the hopper. It was very important for the miller to maintain a constant supply of grain: if the stones 'ran dry' much damage could be done to their carefully prepared surfaces. A special device called a warbler or bell alarm was attached to the hopper to warn the miller if the supply of grain ran low. It consisted of a piece of leather placed near the bottom of the hopper and kept in place by the weight of the grain. A string was attached to the leather and its other end was fixed to a bell. If the grain became depleted the leather pulled out of position allowing the bell to fall against a moving wheel and sound the alarm.

THE STONES

The millwright had to ensure that the heavy millstones stayed in position by providing a strong supporting framework, like that shown on page 13 above, to carry their weight. When the stones were positioned the centre of the runner stone had to coincide exactly with the centre of the stone nut. The stationary bedstone has to be absolutely level and the upper running-stone is very slightly concave. Although the gap between them is paper thin they should never touch — there is a great risk of fire if they do and the surfaces can be badly damaged. The gap between stones which have warmed up while grinding tends to widen, producing unevenly ground flour, and this has to be rectified. So everything has to be carefully balanced.

Dressing stones was a full-time specialist occupation in the days when nearly every parish had a mill of some kind. The very heavy runner stone, which might weigh almost a ton, was lifted from its position on the stationary bedstone and turned over with very simple aids: rope, pulley blocks and large wooden wedges.

Two types of stone were used for grinding. Derbyshire peak stones were made all in one piece and were used for grist. Flour was prepared on French burr stones that were made up from several

ABOVE LEFT: A Peak runner stone.
ABOVE RIGHT: This view of a French stone shows how it was composed of several sections. The bar on which the stone rotates is leaded into the centre.
BELOW: Dressing a bedstone. Here the stone dresser is resting on his left arm and holding the bill in the same hand. The hand used depended upon the situation of the stone. It was not always possible to work around a stone adopting the same position all the time, particularly in the confines of a small post mill. Probably most stone dressers were, by necessity ambidextrous.

sections of stone, cemented together and bound with an iron band. The furrows on a stone's surface were arranged so as to expel the meal around the circumference. The furrows and the flat surfaces (lands) dividing them were made with the hardened steel mill bill. To produce a flat surface on a stone some 4 feet (1.22m) in diameter was a difficult task requiring great skill and precision. When the mill bill struck the stone surface small particles of steel flew off as sparks. Many of these became embedded in the stone-dresser's hands. There is an old tradition that you could identify an experienced man by looking at his hands and getting him to 'show his steel'.

RIGHT: A a stepper, a stepped wedge used in raising runner stones; B a large wooden wedge for supporting upturned runner stones; C mill bills; D thrifts (handle for a bill); E chisel; F file; G mill pick.

BELOW: A millwright's ladle used for pouring molten lead.

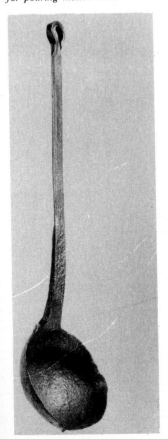

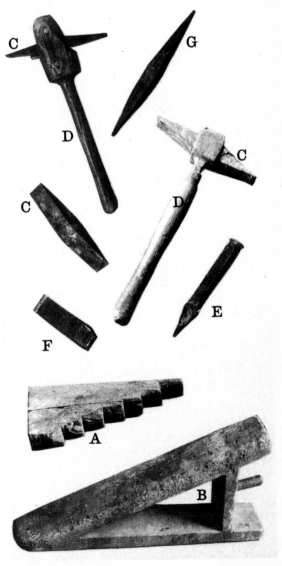

A fine stone tower mill at New Bradwell, Buckinghamshire, which has been restored by the Milton Keynes Development Corporation. The cap resembles an upturned boat. A round tower was a good defence against wind and rain. The slope of the wall is called the 'batter'.

The cap of Copstone smock mill, near Turville, Buckinghamshire, in 1977 after restoration work by Jameson Marshall of Woodbridge.

WINDMILLS

Because they depended on the wind to turn their sails windmills had to be in exposed situations and so were subject to the wear and tear that kept the millwright in business. Windmills need constant care if they are to be kept in order, for the wind and rain quickly undo all the skilful contrivances of the millwright.

Two types of windmill are found in England, the post mill and the tower mill. In the older design, the *post mill*, the bodywork rests on an upright post, upon which it revolves. Below the body (or buck) a system of horizontal and sloping beams forms a trestle to keep the central post in position. Mills of this type had to be turned to keep their sails facing the wind: a tailpole or fan carriage was fixed to the rear of the mill body for this purpose. Millers had to ensure that the sails were always facing the wind, for a strong wind from the rear could blow the mill down.

The inconvenience of turning the entire mill body to align the sails correctly was eliminated some time in the sixteenth or seventeenth century, when an alternative style of mill was introduced. The sails were placed on a revolving cap at the top of the tower. The new *tower mills* had several advantages over post mills: there was more working space available; tower mills were stronger structures than post mills, as they stood firmly on the ground; and although it was still possible for a cap to be blown off or damaged in a severe wind the mill body was usually more resistant to wind and rain. Tower mills built in brick or stone were normally circular or octagonal in plan, with thick sloping walls. Some tower mills were constructed with wooden towers and these were called *smock mills*. Usually hexagonal or octagonal in plan, their exposed corners made it very difficult to make them watertight: wooden towers are very vulnerable at their corners. New techniques and materials, however, can overcome the problem, which was largely insoluble a generation ago.

A drawing of a post mill made about 1815. This elevation shows how the post mill supported the weight of the stones directly on its central post. The design is very interesting as the sloping supports (the quarter bars) that are fixed to the post are duplicated. The mill body is very tall and thin. Working millwrights did not use drawings when they built or repaired a mill. Most of their work was done by traditional methods passed down by word of mouth and by the lessons of experience.

BELOW: This side elevation shows that the brake wheel (the large gear mounted on the sail's axle) drives a lantern wheel like the examples shown on the hand mills. A chain above the steps allowed the miller to operate the brake mechanism via a system of pulleys and levers. Notice how the axle is inclined to allow the sails to tilt backwards, thus missing the bodywork and catching the wind more effectively.

ABOVE: Repairs being made to the post mill at Friston, Suffolk, by Jameson Marshall of Woodbridge in 1976. This mill had a very tall and elaborate ladder fantail mounted on its rear steps. The motion of the fanwheel was transferred to the ground by a series of rods and iron gear wheels.

LEFT: Traditional millwrights did not have powered mobile cranes with an enormous reach like this one being used to remove the mill's cap at Wymondham, Leicestershire. This fine mill is of particular interest as it had six sails. The pulley wheel protruding from the first floor indicates that after the mill lost its sails it was worked for a time by a steam engine.

ABOVE: *After many years of neglect the restoration of the oldest smock mill in England (c. 1650), at Lacey Green, Buckinghamshire, was undertaken by the Chiltern Society. Considerable technical problems faced the voluntary millwrights, such as levelling the uneven curb which supports the cap. This photograph shows the work of protecting the sides. The ancient timbers have been protected by a skin of treated formwork plywood. The layers of Douglas fir are glued together with Recorcinol, which is repellent to woodworm. This skin does not show from the inside as the weatherboarding is placed on the original framework first (see the section above the corrugated iron sheets). When the external cladding of ply is complete a further layer of weatherboarding will be applied to restore the mill's original appearance.*

RIGHT: *A volunteer prepares a section of the new curb to support the cap. The inner radius has to be worked with a concave compass plane. The plane he is using was made by one of his colleagues, and many old millwrights made their own tools too.*

25

An essential part of the millwright's equipment is an extending ladder to help him reach otherwise inaccessible places. The men here are preparing to remove the sailstock from Thorpeness mill, Suffolk.

26

Setting out the new members of the cap frame at Wilton, near Great Bedwyn, Wiltshire, in 1975.

WINDMILL MACHINERY AND SAILS

The power generated by sails has to be transferred from the top of the windmill downwards to the stones. Although this is exactly opposite to the arrangement of a watermill, there are many similarities between watermill and windmill machinery. The pit wheel of a watermill performs the same task as the brake wheel of a windmill; both have a wallower which is mounted on the main shaft.

To repair a windmill's sails the millwright had to work in a very exposed position. The task of removing or replacing heavy members like the sailstocks (which carry the sails) was no easy matter in the days when the principal mechanical aid was a block and tackle. Work on the sails was particularly dangerous in the freezing temperatures and strong winds of the winter months. Considerable skill is required to construct and fit a set of sails that will run evenly. Although the momentum comes from above the transmission of power is the same as in the watermill. The brake wheel mounted on the windshaft meshes with the wallower, as the pit wheel of a watermill does, to turn the main shaft.

27

LEFT: *An adze being used to trim the end of a whip, the main timber of a windmill sail. The adze is one of the oldest of carpenter's tools. Bronze ones were used by the ancient Egyptians, iron ones by the Romans. The Bayeux Tapestry (c. 1072) shows Norman carpenters using them to construct William's invasion fleet.*

BELOW: *The flat head of the adze can also serve as a hammer. Here the cross members of a sailframe are being knocked into their mortises. Each member is set at a different angle. This variation provides the sail with its 'weather' or warp to help it catch the wind. Modern millwrights sometimes use pocket calculators to work out the angles required.*

Wilton mill has two single-shuttered sails and two common sails. Such an arrangement was not unusual in the days of working windmills. The shuttered sails could be adjusted from within the mill without stopping work. The common sails had to be spread with canvas and so grinding time was lost if the sails needed to be reset. At the rear of the domed cap is the fantail which kept the sails facing the wind. The irregular angles of the cross members give the outer edge of the sail a distinct curve, which can be clearly seen here.

29

In March 1976 the common sails at Wilton were spread with canvas for the first time in seventy years, thus demonstrating why the mill needed a stage or gallery surrounding its tower.

This portable corn mill is an example of the ingenuity of Victorian engineers. In the nineteenth century mobile steam engines — drawn by horses in the early days — brought power even to remote farms, enabling the farmer to thresh and grind or even saw timber. It was the village millwright who kept such machines in working order.

OUTWORK

A large part of a millwright's time was spent in corn mills, but there were many other aspects to his work. In East Anglia pumping mills demanded a good deal of attention. Examples of these simple machines with their wooden scoop wheels can still be seen at Herringfleet Marsh, Suffolk, and at Wicken Fen, Cambridgeshire. A restored windpump from Pevensey can now be seen at the Weald and Downland Open Air Museum, Singleton, West Sussex.

In the western counties the heavy crushing pans used in cidermaking were also part of the millwright's work. The large circular stone, which was pulled round by a horse, was also used in other processes such as crushing ore, flint and clinker. With the development of mobile steam engines in the nineteenth century large mobile threshing boxes and even corn mills could be taken to the farms. During the last years of Victoria's reign there were hundreds of contractors who took their tackle from farm to farm at harvest time. The millwright's role expanded to include the repair and maintenance of machinery and the name became a generic term for a mechanic or

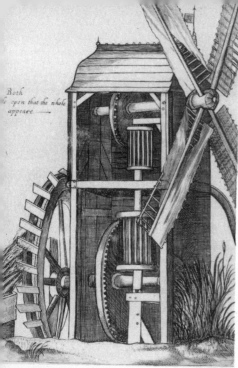

Both
[...] open that the whole
appeare

craftsman thus employed.

The old county directories do not contain very long lists of millwrights: millwrighting work was often carried out by carpenters, wheelwrights and blacksmiths, and millwrights frequently had other occupations, too, like William Sparrow, whose address was the Somerset Wheel and Wagon Works, Martock (1897), and James Webb of the Iron and Brass Foundry, Exning, Suffolk (1892).

Today there is a revived demand for millwrighting skills. Two working millwrights are David Nicholls (Jameson Marshall Ltd), Church House, Erwarton, Ipswich, Suffolk and R. Thompson and Son (J. C. Davies), Alford, Lincolnshire.

ABOVE: A wind-powered water pump illustrated in Walter Blith's 'English Improver Improved' of 1652. Notice the two large lantern pinion wheels.

RIGHT: The fenland of East Anglia was once served by hundreds of wind-driven pumping engines like this one at East Bridge Marsh (which has since been destroyed by the elements). Wooden mills were replaced by the less troublesome steel-framed type of pump seen on the right of the photograph.

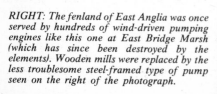

BIBLIOGRAPHY

Bennett, R. and Elton, J. *History of Corn Milling.* 1899.
Freese, Stanley. *Windmills and Millwrighting.* 1957.
Freese, Stanley, and Hopkins, R. T. *In Search of English Windmills.* 1931.
Smith, Arthur. *Windmills in Bedfordshire.* 1975.
Smith, Arthur. *Windmills in Huntingdon and Peterborough.* 1977.
Smith, Arthur. *Windmills in Surrey and Greater London.* 1976.
Smith, Arthur, and Seaby, Wilfred. *Windmills in Warwickshire.* 1977.
Vince, John. *Discovering Watermills.* 1976.
Vince, John. *Discovering Windmills.* 1977.
Vince, John. *Windmills in Buckinghamshire and the Chilterns.* 1977.
Wailes, Rex. *The English Windmill.* 1954.

PLACES TO VISIT

The mills listed here are open to the public at certain times (enquire locally for details). More information about most of them and fuller lists of mills can be found in *Discovering Watermills* and *Discovering Windmills* by John Vince (Shire Publications).

Watermills

Avon: Blaise Castle House Folk Museum, Bristol; Priston Mill, near Bath.
Cornwall: Morden Mill, Cotehele.
Cumbria: Eskdale Mill, Boot; Heron Mill, Beetham, Milnthorpe; Little Salkeld Mill; Muncaster Mill, near Ravenglass.
Devon: Finch Foundry Museum, Sticklepath; Hele Bay Watermill, Ilfracombe; Parracombe Mill.
East Sussex: Park Mill, Bateman's, Burwash; Michelham Priory, Upper Dicker, Hailsham.
Gloucestershire: Arlington Mill, Bibury.
Hampshire: City Mill, Winchester.
Hertfordshire: Kingsbury Watermill Museum, St Albans.
Isle of Wight: Upper Mill, Calbourne; Yafford Mill, Shorwell.
Kent: Chart Mill, Faversham; Crabble Mill, Dover.
Lincolnshire: Alvingham Watermill, near Louth.
Northamptonshire: Billing Mill, Little Billing.
Northumberland: Heatherslaw Mill, Ford.
Somerset: Hornsby Mills, Chard.
South Yorkshire: Abbeydale Industrial Hamlet, Sheffield; Worsbrough Mill Museum, near Barnsley.
Staffordshire: Brindley Mill, Leek; Cheddleton Flint Mill.
Suffolk: Woodbridge Tide Mill.
West Midlands: Sarehole Mill, Moseley, Birmingham.
West Sussex: Woods Mill, Henfield.
Scotland: Preston Mill, East Linton.
Wales: Welsh Folk Museum, St Fagans, near Cardiff.

Windmills

Buckinghamshire: Lacey Green; New Bradwell; Pitstone Green; Quainton.
East Sussex: Argos Hill, Mayfield; Nutley; Polegate.
Essex: Bocking; Stansted Mountfitchet.
Hereford and Worcester: Avoncroft Museum of Buildings, near Bromsgrove.
Humberside: Wrawby.
Isle of Wight: Bembridge.
Kent: Chillenden; Draper's Mill, Margate; Herne; Meopham Green; Union Mill, Cranbrook.
Lincolnshire: Alford; Burgh-le-Marsh; Heckington.
Norfolk: Berney Arms, near Great Yarmouth; Billingford; Horsey Drainage Mill; Sutton, near Stalham.
Somerset: High Ham.
Suffolk: Friston; Herringfleet; Holton; Saxtead Green.
Surrey: Outwood; Reigate Heath.
West Midlands: Berkswell.
West Sussex: Belloc's Mill, Shipley; Salvington Mill, near Worthing; Weald and Downland Open Air Museum, Singleton.
Wiltshire: Wilton, near Great Bedwyn.